YOUR KNOWLEDGE HAS VALUE

- We will publish your bachelor's and
  master's thesis, essays and papers

- Your own eBook and book -
  sold worldwide in all relevant shops

- Earn money with each sale

Upload your text at www.GRIN.com
and publish for free

Imprint:

Copyright © 2014 GRIN Verlag, Open Publishing GmbH
Print and binding: Books on Demand GmbH, Norderstedt Germany
ISBN: 978-3-668-11240-7

Yaw Boateng

Phytoremediation of Heavy Metal Contaminated Soil Using Leucaena Leucocephala

A Case Study at Anglogold Ashanti Obuasi Ghana

GRIN Publishing

THESIS PRESENTATION

TOPIC:

PHYTOREMEDIATION OF HEAVY METAL CONTAMINATED SOIL USING *Leucaena leucocephala.* A case study at AngloGold Ashanti Ghana

BY

BOATENG YAW

INTRODUCTION

Background of Study

- Economic, agricultural and industrial developments are often linked to polluting the environment which include environmental soils.

- Sources of metal enrichment of soil include incinerators, fertilizers, urban compost, car exhausts, cement factories, residues from mining and smelting industries, sludge and sewage.

- Heavy metals are usually associated with pollution and toxicity although some of these elements (essential metals) are required by organisms at low concentrations (Adriano, 2001).

INTRODUCTION cont'd

- Elements associated with gold mining waste includes arsenic(As), cadmium (Cd), copper (Cu), lead (Pb), antimony (Sb) and zinc (Zn) (Ferreira da Silva *et al.*, 2004).

- Phytoremediation involves using plants and their associated micro biota to solve pollution problems and in this case metal polluted soils.

INTRODUCTION cont'd

Justification

- Accumulation of heavy metals in the environment pose a threat to both human health and the natural environment.

- Conventional remediation technologies (solidification and stabilization, soil flushing, electrokinetics, chemical reduction/oxidation, soil washing, low temperature thermal desorption, incineration, vitrification, pneumatic fracturing, excavation/retrieval and landfill disposal) are expensive and destructive (Mellem, 2008).

- In places where the heavy metal contaminated waste are contained, according to Renault *et. al* (2005) wind and water can physically move tailings off-site causing contamination of adjacent areas.

INTRODUCTION cont'd

- Phytoremediation appears as a valid option since it is best suited for the remediation of these diffusely polluted areas and at much lower costs than other methods (Kumar *et al.*, 1995).

- *Lucaena leucocephala* is a nitrogen fixing plant which will further add nitrogen to the soil .

- This study therefore seeks to contribute to the search for plants for phytoremediation.

INTRODUCTION cont'd

Aim and Objectives

The aim is to determine the capability of *Leucaena leucocephala* in phytoremediation of heavy metal contaminated soils.

Specific Objectives

- To determine the levels of heavy metals accumulation in the *Leucaena leucocephala*.
- To determine the effect of inorganic fertilizer (NPK) on heavy metal accumulation by the *Leucaena leucocephala*.
- To determine the effect of organic manure (PKC) on enhancing phytoremediation of heavy metals by *Leucaena leucocephala*.
- To determine the potential of *Leucaena leucocephala* as a hyperaccumulator for specific heavy metals

MATERIALS AND METHODS

Study Site

Nursery and revegetation unit of AngloGold Ashanti in Obuasi

Collection of Tailings Soil Samples

Eastern part of the Sansu tailings dam. An area of $40m^2$ was divided into 8 equal zones and further divided into 5 subzones where 20kg of tailing soil was collected from each subzone at a depth of 30cm with soil auger. 100kg of tailing soil was collected from each zone making a total 800kg of tailing soil collected in sacks.

MATERIALS AND METHODS cont'd

Control soil was obtained from Mampanhwe. An area of $20m^2$ was selected and divided into 5 equal zones with each zone having an area of $4m^2$. 6 spots were then randomly selected from each zone and 10kg of soil was collected from each spot. The soil was taking at a depth of 40cm and a total of 300kg of top soil was taking as control.

Collection of Planting Material

Seeds were collected from the tailings dam at a distance of 130m from the dam.

Nursing and Transplanting

Nursery beds were watered each morning. Seedlings were nursed for 3 weeks at the nursery. After 3 weeks they were transplanted.

MATERIALS AND METHODS cont'd

Experimental Design

- Layout of the experiment was Randomised Complete Block Design.
- 120 poly-pots of size 8 x 10 inches were filled with 5kg of treatment soil.
- 10 treatments with each treatment replicated 6 times for the two harvest periods.

Treatments used

Treatment 1- T1 (Tailings soil alone)

Treatment 2 - T2 (Tailing soil + chelator (EDTA))

MATERIALS AND METHODS cont'd

Treatment 3 - T3 (Tailing soil + Fertilizer (NPK))

Treatment 4 - T4 (Tailing soil + Fertilizer + Chelator (EDTA))

Treatment 5 - T5 (Tailing soil + Palm kernel Cake)

Treatment 6 - T6 (Tailing soil + Palm kernel Cake + Chelator (EDTA))

Treatment 7 - T7 (Tailing soil + Topsoil) (3:2)

Treatment 8 - T8 (Tailing soil + Topsoil) (2:3)

Treatment 9 - T9 (Tailing soil + Topsoil) (1:1)

Treatment 10 - T10 (Topsoil or Control)

MATERIALS AND METHODS cont'd

- Chelator (EDTA) was prepared by dissolving 60g of EDTA salt in 500ml of distilled water. The concentration used was 0.3M of which 25 ml was added a week before harvesting to prevent loss of shoots which might be concentrated with lead. According to Larson *et al.*, (2007), chelates should be applied within one week of treatment to avoid loss of shoots.

- Treatment with inorganic fertilizer (NPK) was prepared by dissolving 370g (equivalent to 2 full milk tins of NPK) in 6 litres of water of which 150ml was mixed with tailings.

- Treatment with organic manure (PKC) was prepared by mixing 5kg of tailing soil with 120g of palm kernel cake (PKC).

MATERIALS AND METHODS cont'd

Harvesting

- First harvest was done 45 days after transplanting and 7 days after EDTA application.

- Samples were washed with distilled water and separated into above shoots and roots.

- The final harvest was done 30 days after the first harvest.

- Treatment soils were analysed for pH and heavy metals present after each harvest.

MATERIALS AND METHODS cont'd

Soil analysis

- NPK and particle size determination for tailing soil and control soil

- pH and heavy metal (As, Fe, Pb, Zn, Cd and Cu) analysis were done for treatments before transplanting, after first harvest and after second harvest.

Plant Analysis

- Heavy metal content in shoots and roots were determined before transplanting, after first harvest and after second harvest.

- Fresh and dry weights in whole plant were determined after first and second harvest.

MATERIALS AND METHODS cont'd

Accumulation ratios

Was used to compare performance of various treatment plants

$$AR = \frac{Level\ of\ heavy\ metals\ in\ treated\ plant\ part\ at\ harvest}{Level\ of\ heavy\ metals\ in\ plant\ part\ before\ transplanting}$$

Bioaccumulation ratios

Was calculated by using the procedure described by Cai and Lena (2003) and expressed as:

$$BR = \frac{Accumulation\ in\ Plant}{Accumulation\ in\ soil}$$

MATERIALS AND METHODS cont'd

Translocation Ratio (TF)

Was used to determine if metals were translocated from the roots to the aerial parts of the plant.

$$TF = \frac{Metal\ concentration\ in\ shoot}{Metal\ concentration\ in\ root}$$

Percentage Reduction

$$PR = \frac{Metal\ level\ before\ transplanting - Metal\ level\ after\ first\ harvest}{Metal\ level\ before\ transplanting}) \times 100$$

RESULTS AND DISCUSSION

Physichochemical parameters of soil

According to Suttie (2005) *L. leucocephala* does best on deep, well drained, neutral to calcareous soils. However, it grows on a wide variety of soil types including mildly acid soils.

Low nutrient in tailings accounts for the slow growth observed

Table 1. Physichochemical parameters of soil

Parameter	Tailing soil	Top Soil (Control Soil)
pH	7.13±0.58	5.66±0.30
Total N (%)	0.08±0.02	0.13±0.03
Available P (mg/g)	0.02±0.01	0.45±0.25
Available K (mg/g)	0.03±0.01	0.36±0.03
Sand %	20.21±2.02	24.00±0.35
Clay %	5.46±0.03	53.75±2.50
Silt %	75.25±1.51	28.25±0.60
Soil type	Silty sand	Clayey loam

RESULTS AND DISCUSSION cont'd
Table 2. Levels of heavy metals in treatment soils before transplanting

Trt	As	Fe	Pb	Zn	Cd	Cu
T1	910.67±14.24[h]	4261.69±96.57[f]	38.04±2.31[g]	100.79±1.37[cd]	10.60±0.39[d]	71.91±1.19[d]
T2	901.11±8.26[gh]	4392.09±32.99[g]	33.21±1.72[e]	103.79±2.85[de]	11.08±0.31[de]	70.61±1.22[cd]
T3	898.96±2.24[g]	4618.45±108.88[h]	35.65±0.50[f]	103.03±1.15[cde]	11.47±0.56[e]	74.00±2.69[de]
T4	896.52±5.12[fg]	4203.08±59.46[f]	32.58±0.52[de]	104.51±2.83[e]	10.67±0.34[d]	70.66±1.61[cd]
T5	884.33±4.61[e]	3685.55±46.97[e]	30.95±1.21[cd]	102.51±1.11[cde]	11.18±0.30[de]	75.91±0.82[e]
T6	886.96±6.16[ef]	3477.82±19.33[d]	31.07±0.89[cd]	101.09±1.34[cd]	10.90±0.09[de]	74.91±2.13[e]
T7	870.26±5.40[d]	3243.56±23.73[c]	26.96±0.36[b]	100.14±0.98[bc]	9.69±0.33[c]	67.09±1.14[c]
T8	755.54±3.87[b]	2911.30±51.19[b]	26.36±0.54[b]	97.77±1.03[b]	8.25±0.18[b]	56.05±2.37[b]
T9	797.77±4.74[c]	3477.25±51.62[d]	29.33±0.80[c]	100.51±1.65[bc]	9.41±0.41[c]	59.09±4.86[b]
T10	280.09±5.20[a]	1323.19±9.76[a]	23.82±0.63[a]	89.42±1.47[a]	6.73±0.29[a]	29.39±0.55[a]

Mean ± SD in the same column with different letters differ significantly ($p < 0.05$).

RESULTS AND DISCUSSION cont'd

- As in all the treatment soils exceeded the normal concentrations allowed due the underground arsenic bearing rock called Arsenopyrite.

- Iron (Fe), Lead (Pb) and Cd exceeded the normal concentrations values in soils in all treatments.

- Zinc concentration was generally below the normal values in soils according to the European Union Regulatory Standards.

- Copper (Cu) the normal values allowed in soils according to European Union Regulatory Standards was exceeded by treatments that contained the tailings soil (T1 to T9).

RESULTS AND DISCUSSION cont'd

Fig. 1 Levels of metals in plant shoot and roots before transplanting

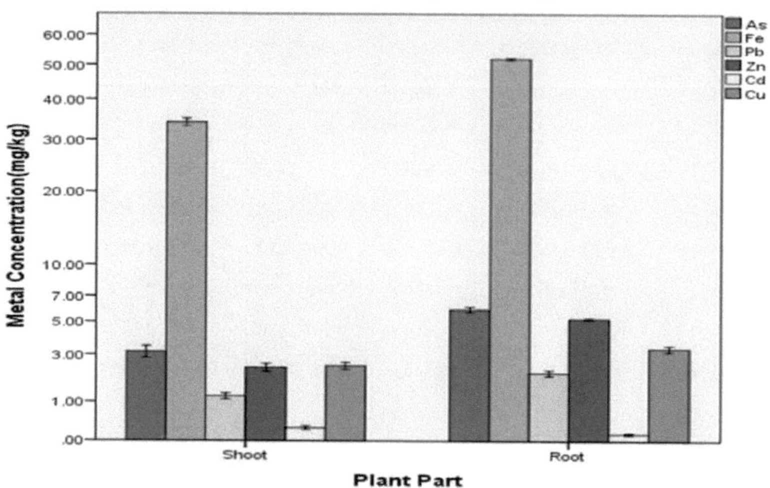

RESULTS AND DISCUSSION cont'd

Table 3. Metal concentration in shoots at 1st harvest

Treatment	As	Fe	Pb	Zn	Cd	Cu
T1	38.08±1.89 f	52.20±1.78 c	1.56±0.19 b	7.67±0.43 c	0.87±0.03 ef	2.92±0.40 e
T2	71.36±1.10 h	132.94±1.32 i	2.26±0.14 f	10.08±0.46 f	0.94±0.05 f	5.77±0.09 f
T3	31.98±1.03 e	59.64±1.03 d	1.82±0.15 cd	5.79±0.15 b	0.67±0.02 c	1.82±0.01 b
T4	43.18±0.44 g	89.29±0.56 h	2.14±0.12 ef	8.84±0.51 e	0.86±0.04 e	3.01±0.08 e
T5	21.81±1.06 c	142.61±0.53 j	1.13±0.05 a	3.61±0.32 a	0.41±0.04 a	1.72±0.02 b
T6	24.84±1.10 d	87.13±0.94 g	2.51±0.26 g	7.43±0.17 c	0.91±0.03 ef	2.48±0.16 d
T7	9.91±0.51 a	34.98±0.77 b	1.25±0.12 a	8.35±0.10 d	0.52±0.02 b	1.12±0.04 a
T8	11.42±0.49 a	31.24±1.18 a	1.36±0.20 ab	7.60±0.15 c	0.41±0.07 a	1.30±0.04 a
T9	44.13±0.62 g	64.42±0.81 e	1.92±0.05 de	7.44±0.11 c	0.67±0.03 c	1.68±0.07 b
T10	14.09±1.58 b	76.51±1.05 f	1.60±0.12 bc	7.28±0.18 c	0.78±0.05 d	2.12±0.13 c

Mean ± SD in the same column with different letters differ significantly (p < 0.05).

Table 4. Metal concentration in shoots at 2nd harvest

Treatment	As	Fe	Pb	Zn	Cd	Cu
T1	66.61±1.44 fg	84.77±2.71 b	2.64±0.32 b	11.30±2.13 c	1.33±0.23 d	4.09±0.83 cd
T2	103.90±1.53 h	192.26±0.00 g	3.83±0.08 c	15.08±1.12 d	1.67±0.16 e	8.32±0.45 e
T3	44.89±3.94 e	84.40±3.37 b	2.53±0.20 b	8.09±0.17 b	0.98±0.09 c	3.43±0.67 cd
T4	68.03±4.47 g	137.78±2.46 f	3.68±0.15 c	12.10±0.93 c	1.38±0.30 de	4.22±0.14 d
T5	31.24±2.86 c	190.45±12.69 g	1.59±0.28 a	5.72±1.30 a	0.64±0.12 ab	2.61±0.08 b
T6	35.45±3.28 d	116.64±5.84 d	3.78±0.02 c	10.68±1.92 c	1.53±0.26 de	3.68±0.44 cd
T7	14.12±1.16 a	50.89±1.30 a	2.36±0.32 b	12.34±0.12 c	0.86±0.07 bc	1.71±0.24 a
T8	17.80±0.46 a	44.88±0.56 a	2.56±0.46 b	10.91±0.73 c	0.50±0.10 a	2.08±0.10 ab
T9	63.37±0.10 f	96.03±4.07 c	3.87±0.38 c	11.32±0.89 c	1.26±0.14 d	3.36±0.62 c
T10	26.52±1.09 b	127.35±7.58 e	3.43±0.70 c	10.91±0.39 c	1.38±0.17 de	3.45±0.03 d

Mean ± SD in the same column with different letters differ significantly (p < 0.05).

RESULTS AND DISCUSSION cont'd

Table 5. Accumulation ratios in shoots at 1st harvest

Treatment	Metals (mg/kg)					
	As	Fe	Pb	Zn	Cd	Cu
T1	6.21	1.53	1.30	3.33	3.11	1.19
T2	22.58	3.90	1.88	4.38	3.36	2.36
T3	10.12	1.75	1.52	2.52	2.39	0.74
T4	13.66	2.62	1.78	3.84	3.07	1.23
T5	6.90	4.19	0.94	1.57	1.46	0.70
T6	7.86	2.56	2.09	3.23	3.25	1.01
T7	3.13	1.03	1.04	3.63	1.86	0.46
T8	3.61	0.92	1.13	3.30	1.46	0.53
T9	13.97	1.89	1.60	3.23	2.39	0.69
T10(control)	4.46	2.25	1.33	3.17	2.79	0.87

Table 6. Accumulation ratios in shoots at 2nd harvest

Treatment	Metals (mg/kg)					
	As	Fe	Pb	Zn	Cd	Cu
T1	21.08	2.49	2.20	4.91	4.75	1.67
T2	32.91	5.64	3.19	6.56	5.96	3.40
T3	14.21	2.48	2.11	3.52	3.50	1.40
T4	21.53	4.05	3.07	5.26	4.93	1.72
T5	9.89	5.59	1.33	2.49	2.29	1.07
T6	11.22	3.42	3.15	4.64	5.46	1.50
T7	4.47	1.49	1.97	5.37	3.07	0.70
T8	5.63	1.32	2.13	4.74	1.79	0.85
T9	20.05	2.82	3.23	4.92	4.50	1.37
T10	8.39	3.74	2.86	4.74	4.93	1.41

RESULTS AND DISCUSSION cont'd

- Generally the levels of metal concentrations and accumulation ratios in all the treatments increased from first to second harvest.

- The highest accumulation ratio recorded in T2 deviated from Dias *et al.* (2009) work which observed that As levels were relatively low in young leaves of *L. leucocephala*.

- The low accumulation ratio for Fe in T8 and Pb in T5 at the first harvest could be as a result of the fact that the treatment could not support their accumulations in the shoot earlier.

- The low accumulation of Cu for T3, T5, T7, T8, T9 and T10 for the first harvest and T7 and T8 for the second harvest could be due to its slow Cu accumulation in shoots.

RESULTS AND DISCUSSION cont'd

Table 7. Metal concentration in roots at 1st harvest

Trt	As	Fe	Pb	Zn	Cd	Cu
T1	66.62±0.64 [g]	265.45±1.09 [h]	9.45±0.32 [e]	27.05±1.18 [a]	2.50±0.16 [fg]	25.99±1.42 [f]
T2	126.26±0.87 [i]	618.42±0.69 [i]	9.57±0.47 [e]	53.70±0.53 [h]	1.45±0.16 [cd]	17.71±0.17 [d]
T3	34.32±0.50 [c]	484.63±1.42 [h]	10.81±0.25 [f]	10.70±0.11 [c]	1.44±0.16 [cd]	10.55±0.35 [b]
T4	35.64±0.68 [d]	462.62±2.66 [f]	5.62±0.15 [b]	45.31±0.43 [g]	1.67±0.17 [de]	17.65±0.15 [d]
T5	212.06±0.88 [j]	477.19±3.94 [g]	12.63±0.13 [g]	11.07±0.05 [c]	0.76±0.07 [b]	18.70±0.05 [e]
T6	79.76±0.77 [h]	428.82±4.74 [e]	10.62±0.14 [f]	18.04±0.07 [d]	2.60±0.09 [g]	28.15±0.50 [h]
T7	44.60±0.53 [f]	331.52±7.04 [c]	6.61±0.12 [c]	39.46±0.66 [f]	1.85±0.10 [e]	27.00±0.41 [g]
T8	26.83±0.54 [b]	151.50±1.23 [a]	3.60±0.21 [a]	10.50±0.18 [c]	0.51±0.24 [a]	17.51±0.29 [d]
T9	41.90±0.12 [e]	267.94±1.29 [h]	7.87±0.07 [d]	9.73±0.13 [b]	2.33±0.16 [f]	14.32±0.16 [c]
T10	19.89±0.29 [a]	421.10±1.31 [d]	9.10±0.72 [e]	3.34±0.29 [a]	1.31±0.23 [c]	9.20±0.31 [a]

Mean ±SD in the same column with different letters differ significantly (p < 0.05).

Table 8. Metal concentration in roots at 2nd harvest

Trt	As	Fe	Pb	Zn	Cd	Cu
T1	109.79±10.47 [e]	381.83±0.72 [b]	15.71±1.05 [d]	39.93±0.43 [d]	3.64±0.33 [e]	39.45±0.88 [d]
T2	182.99±7.85 [f]	983.05±40.85 [f]	14.59±1.80 [d]	75.38±5.20 [g]	2.40±0.14 [c]	28.59±1.39 [c]
T3	49.96±1.44 [bc]	679.38±3.80 [e]	14.61±1.95 [d]	14.12±0.81 [b]	2.51±0.32 [cd]	14.56±0.23 [a]
T4	56.96±0.76 [cd]	680.81±0.66 [e]	7.54±0.30 [b]	64.29±3.28 [f]	2.14±0.41 [c]	27.34±3.30 [c]
T5	304.76±11.87 [g]	642.30±8.19 [e]	17.73±1.66 [e]	15.16±0.65 [b]	1.39±0.26 [b]	27.32±2.16 [c]
T6	102.02±0.48 [e]	600.69±55.27 [d]	14.55±1.79 [d]	27.47±1.01 [c]	3.91±0.06 [e]	41.80±3.89 [d]
T7	61.71±0.49 [d]	462.14±18.47 [c]	9.46±0.34 [bc]	54.76±2.73 [e]	2.88±0.15 [d]	40.80±2.92 [d]
T8	41.75±4.03 [b]	226.60±20.60 [a]	5.22±0.13 [a]	14.29±0.13 [b]	0.74±0.05 [a]	25.17±0.19 [c]
T9	64.24±3.78 [d]	396.48±15.04 [b]	11.04±0.85 [c]	14.96±0.40 [b]	3.52±0.25 [e]	21.51±1.93 [b]
T10	32.06±0.46 [a]	652.79±20.28 [e]	13.92±1.24 [d]	4.77±0.54 [a]	2.43±0.45 [cd]	16.30±0.07 [a]

Mean ±SD in the same column with different letters differ significantly (p < 0.05).

RESULTS AND DISCUSSION cont'd

Tabe 9. Accumulation ratio in roots at 1st harvest

Treatment	Metals (mg/kg)					
	As	Fe	Pb	Zn	Cd	Cu
T1	11.23	5.10	4.44	5.16	17.86	7.76
T2	21.29	11.89	4.49	10.25	10.36	5.29
T3	5.79	9.32	5.08	2.04	10.29	3.15
T4	6.01	8.89	2.64	8.65	11.93	5.27
T5	35.76	9.17	5.93	2.11	5.43	5.58
T6	13.45	8.24	4.99	3.44	18.57	8.40
T7	7.52	6.37	3.10	7.53	13.21	8.06
T8	4.53	2.91	1.69	2.00	3.64	5.23
T9	7.07	5.15	3.69	1.86	16.64	4.27
T10	3.35	8.09	4.27	0.64	9.36	2.75

Table 10. Accumulation ratio in roots at 2nd harvest

Treatment	Metals (mg/kg)					
	As	Fe	Pb	Zn	Cd	Cu
T1	18.51	7.34	7.38	7.62	26.00	11.78
T2	30.86	18.90	6.85	14.39	17.14	8.53
T3	8.42	13.06	6.86	2.69	17.93	4.35
T4	9.61	13.09	3.54	12.27	15.29	8.16
T5	51.39	12.35	8.32	2.89	9.93	8.16
T6	17.20	11.55	6.83	5.24	27.93	12.48
T7	10.41	8.88	4.44	10.45	20.57	12.18
T8	7.04	4.36	2.45	2.73	5.29	7.51
T9	10.83	7.62	5.18	2.85	25.14	6.42
T10	5.41	12.55	6.54	0.91	17.36	4.87

RESULTS AND DISCUSSION cont'd

- The plant generally accumulated the metals in roots than the shoots.

- According to Dias *et al.* (2009), the highest As concentration in *L. Leucocephala* are mostly found in the roots and the study attested to that fact.

- The study confirms the proposition made by Saraswat and Rai (2011) that the plant accumulates Zn and Cd mostly in its roots.

RESULTS AND DISCUSSION cont'd

Table 11. Metals accumulated in whole plant at 1st harvest

Trt	Metals (mg/kg)					
	As	Fe	Pb	Zn	Cd	Cu
T1	11.52	3.69	3.34	4.61	8.02	4.98
T2	21.74	8.73	3.58	8.46	5.69	4.05
T3	7.29	6.32	3.83	2.19	5.00	2.13
T4	8.67	6.41	2.35	7.18	6.02	3.56
T5	25.73	7.20	4.17	1.95	2.79	3.52
T6	11.51	5.99	3.98	3.38	8.36	5.28
T7	6.00	4.26	2.38	6.34	5.64	4.85
T8	4.21	2.12	1.50	2.40	2.19	3.24
T9	9.46	3.86	2.97	2.28	7.14	2.76
T10	3.74	5.78	3.24	1.41	4.98	1.95

Table 12. Metals accumulated in whole plant at 2nd harvest

Trt	Metals (mg/kg)					
	As	Fe	Pb	Zn	Cd	Cu
T1	19.41	5.42	5.56	6.79	11.86	7.51
T2	31.57	13.65	5.58	12.00	9.69	6.36
T3	10.43	8.87	5.20	2.95	8.31	3.10
T4	13.75	9.51	3.40	10.13	8.38	5.44
T5	36.96	9.67	5.85	2.77	4.83	5.16
T6	15.12	8.33	5.55	5.06	12.95	7.84
T7	8.34	5.96	3.58	8.90	8.90	7.33
T8	6.55	3.15	2.36	3.34	2.95	4.70
T9	14.04	5.72	4.52	3.48	11.36	4.29
T10	6.44	9.06	5.25	2.08	9.07	3.41

RESULTS AND DISCUSSION cont'd

Table 13. Translocation factor at 1st harvest

Trt	As	Fe	Pb	Zn	Cd	Cu
T1	0.57	0.20	0.17	0.28	0.35	0.11
T2	0.57	0.22	0.24	0.18	0.65	0.33
T3	0.94	0.12	0.17	0.54	0.47	0.18
T4	1.21	0.19	0.38	0.20	0.52	0.17
T5	0.10	0.30	0.09	0.33	0.54	0.09
T6	0.31	0.20	0.24	0.41	0.35	0.09
T7	0.22	0.11	0.19	0.21	0.28	0.04
T8	0.43	0.21	0.38	0.73	0.85	0.07
T9	0.57	0.20	0.17	0.28	0.35	0.11
T10	0.57	0.22	0.24	0.18	0.65	0.33

Table 13. Translocation factor at 2nd harvest

Trt	As	Fe	Pb	Zn	Cd	Cu
T1	0.61	0.23	0.17	0.29	0.38	0.11
T2	0.57	0.20	0.27	0.22	0.68	0.29
T3	0.90	0.14	0.18	0.57	0.40	0.24
T4	1.20	0.21	0.49	0.19	0.69	0.16
T5	0.11	0.30	0.09	0.38	0.46	0.10
T6	0.35	0.20	0.26	0.39	0.39	0.10
T7	0.24	0.11	0.26	0.23	0.30	0.04
T8	0.43	0.20	0.50	0.45	0.67	0.08
T9	1	0.25	0.36	0.76	0.36	0.16
T10	0.83	0.22	0.26	2.32	0.60	0.21

RESULTS AND DISCUSSION cont'd

- Generally the plant was not translocating most metals to the shoots including those with chelator (T2 and T6).

- But it maintained high concentrations in the roots of most of the treatments. It can therefore be employed in regenerating heavy contaminated soils according to Baker (1981).

- Since the translocation ratios increased from first to second harvest, they could have increased if the duration for study was longer.

RESULTS AND DISCUSSION cont'd

Table 14. Bioaccumulation factor at 1st harvest

Trt	As	Fe	Pb	Zn	Cd	Cu
T1	0.12	0.08	0.28	0.35	0.32	0.41
T2	0.22	0.17	0.37	0.62	0.22	0.33
T3	0.07	0.12	0.36	0.16	0.18	0.17
T4	0.09	0.13	0.24	0.51	0.24	0.30
T5	0.26	0.17	0.46	0.14	0.11	0.27
T6	0.12	0.15	0.43	0.26	0.33	0.42
T7	0.06	0.11	0.30	0.48	0.25	0.42
T8	0.05	0.06	0.20	0.19	0.12	0.36
T9	0.11	0.09	0.34	0.17	0.31	0.27
T10	0.13	0.38	0.45	0.12	0.31	0.39

Table 15. Bioaccumulation factor at 2nd harvest

Trt	As	Fe	Pb	Zn	Cd	Cu
T1	0.20	0.11	0.47	0.52	0.47	0.61
T2	0.32	0.27	0.57	0.87	0.36	0.52
T3	0.11	0.17	0.48	0.22	0.30	0.24
T4	0.14	0.20	0.35	0.74	0.32	0.45
T5	0.38	0.23	0.63	0.20	0.18	0.39
T6	0.16	0.21	0.60	0.38	0.50	0.62
T7	0.09	0.16	0.44	0.68	0.39	0.64
T8	0.08	0.10	0.30	0.21	0.15	0.48
T9	0.16	0.14	0.50	0.27	0.51	0.42
T10	0.21	0.60	0.72	0.18	0.58	0.68

RESULTS AND DISCUSSION cont'd

- Based on the duration and conditions provided for the study due to the bioaccumulation ratios, the plant can be said to be a poor hyperaccumulator of heavy metals.

- Gardezi *et al.* (2008) indicated that work done with *L. leucocephala* should be given a minimum of 1 year to allow the plant to mature so that it can reach its bioaccumulation capacity.

- Though the plant did not perform well based on its bioaccumulation factor, it was able to tolerate the tailing soil which is highly polluted with heavy metals and accumulated some biomass.

RESULTS AND DISCUSSION cont'd

Effect of Fertilizer (NPK) on Metal Concentration in Plants

- The effect of the fertilizer in increasing the biomass of the plant was not seen when compared to the control. The same incident of low biomass was also experienced by Aziz (2011).

- The accumulation ratios were all greater than 1 (>1).

- T4 performed better when it comes to the number of metals it was able to accumulate than T3 in the whole plant which could be due to the chelator.

- The biomass could have probably highly increased during the second harvest if fertilizer application was not one time and may be, the quantity varied.

RESULTS AND DISCUSSION cont'd

Effect of Chelator on Metal Concentration in Plants

- The accumulation ratios in all the treatments that had chelator added were greater than 1 (>1) in both harvests.

- Arsenic (As) was having the highest accumulation ratios in all the treatments for both harvests but the concentration was mostly in the roots except in T4.

- This could be attributed to the time of adding the chelator so if it had been added earlier it could have aided in translocating the metal to the shoots and also biomass was not enough.

- But because of the toxic effects, it is recommended that chelates should be applied only after a maximum amount of plant biomass has been produced and prompt harvesting (within one week of treatment) is required to minimize the loss of Pb-laden shoots (Larson *et al.*, 2007; Aziz, 2011).

RESULTS AND DISCUSSION cont'd

Effect of Palm Kernel Cake on Metal Concentration in Plants

- The idea behind the addition of the PKC was to help the plant to increase its biomass and also accumulation of the metals but the biomass obtained was less than the control.

- It could be due to the one time application and also the quantity which might not be enough. According to Kolade *et al.* (2005) PKC should be converted into compost and applied 4t/ha to obtain yields comparable to those of organo-minerals fertilizer and chemical fertilizers.

- The accumulation ratios for the treatments that contained PKC for both harvests in the whole plant were greater than 1 (>1). T5 was able to accumulate As, Fe and Pb higher than T6 which was able to accumulate Zn, Cd and Cu.

CONCLUSION AND RECOMMENDATION

Conclusion

- To survive high concentrations of heavy metals in soils, plants can either stabilize metal contaminants in the soil through avoidance or can take up the contaminants into their cellular structure by tolerating them as was described by other researchers. And so far as the plant has been able to grow and accumulate biomass and tolerate these metals, it has the capacity to take up the metals as well as tolerate the stress they gave it.

- The plant could also be seen to have slow growth due to the conditions that it found itself but when if the duration of study had increased it could have performed wonderfully. Gardezi *et al.* (2008) indicated that work done with *L. Leucocephala* should be given a minimum of 1 year to allow the plant to mature so that it can reach its bioaccumulation capacity

CONCLUSION AND RECOMMENDATION

Recommendations

* Duration of the study was not enough and future studies should consider at least one year as it has been talked about by other researchers.

* Chelator such as EDTA should be used for such a study by varying the quantity and also the time of addition to see its effect.

* Fertilizer (NPK) and PKC were added at once during the treatment preparation stage and this did not help in accumulating enough biomass which could be due to the quantities added or the one time application so future further studies should vary their quantities and also further subsequent additions with time.

REFERENCES

* **Adriano**, D.C. (2001). Cadmium. In Adriano D.C. (Ed.), Trace elements in terrestrial environments, biogeochemistry, bioavailability, and risks of metals. 2nd edition, Springer-Verlag, New York, pp. 264-314.

* **Ferreira da Silva,** E., Zhang, C., Serrano Pinto, L., Patinha, C. and Reis, P. (2004). Hazard assessment on arsenic and lead in soils of Castromil gold mining area, Portugal. Applied Geochemistry, 19(6): 887-898

* **Mellem**, J.J. (2010). Phytoremediation of heavy metals using *Amaranthus dubius*, Theses submitted to Department of Biotechnology and Food Technology, Durban University of Technology, Durban, South Africa.

* **Renault**, S. and Green, S. (2005). Phytoremediation and revegetation of mine tailings and bio-ore production: effects of paper mill sludge on plant growth in tailings from Central Manitoba (Au) minesite (NTS 52L13); *In:* Report of Activities 2005, Manitoba Industry, Economic Development and Mines, Manitoba Geological Survey, p. 167–169.

* **Larson**, S.L., Teeter, C. L., Medina, V.F. and Martin, W.A. (2007). Environmental Quality and Technology Program Treatment and Management of Closed or Inactive Small Arms Firing Ranges. US Army

REFERENCES

Kumar, N., Dushenkov, V., Motto, H. and Raskin, I. (1995). Phytoextraction: the use of plants to remove heavy metals from soils. Environmental Science and Technology 29.

Sutie, J.M. (2005). Food and Agriculture Organization (FAO), *Leucaena leucocephala* (Lam.) de Wit. http://www.fao.org/ag/AGP/AGPC/doc/Gbase/DATA/Pf000158.htm 23/06/2005 10:34:22

Dias, L.E., Melo, R.E., Vargas de Melo, J.W., Oliveira, J.A. and Daniels, W.L. (2010). Potential of three legume species for phytoremediation of Arsenic contaminated soils.Soil Dep. Universidade Federal de Viçosa – UFV, 36571-000, Viçosa-MG Brazil.

Saraswat S. and Rai J.P. (2011). Prospective application of *Leucaena leucocephala* for phytoextraction of Cd and Zn and nitrogen fixation in metal polluted soils. Ecotechnology Laboratory, Department of Environmental Sciences, G.B. Pant University of Agriculture & Technology, Pantnagar, India

Baker, A.J.M. (1981). Accumulators and excluders – Strategies in the response of plants to heavy metals, J. Plant Nutr. 3(1-4): 643-654.

Aziz, F. (2011). Phytoremediation of Heavy Metal Contaminated Soil Using *Chromolaena odorata* and *Lantana camara*. Master's Thesis for the Award of MSc. Environmental Science. Department of Theoretical and Applied Biology, KNUST Kumasi.

Gardezi, A.K., Barceló, I.D., García, A.E., Saenz, E.M., Saavedra, U.L., Sergio R., Márquez, B., Verduzco, C.E., Gardezi, H. and Talevera-Magaña, D. T. (2008). Cu^{2+} Bioaccumulation by *Leucaena leucocephala* in symbiosis with *Glomus* spp. and *Rhizobium* in Copper-containing soil.Colegio de Postgraduados. Instituto de Recursos Naturales, Programa Hidrociencias. Montecillo, Texcoco, Edo. de México

THANK YOU

YOUR KNOWLEDGE HAS VALUE

- We will publish your bachelor's and master's thesis, essays and papers

- Your own eBook and book - sold worldwide in all relevant shops

- Earn money with each sale

Upload your text at www.GRIN.com and publish for free